My Scientific Letter to NASA, SpaceX and Blue Origin

BALUNGI FRANCIS

Copyright © Balungi Francis 2020
Copyright © Bill Stone Services 2020

Balungi Francis asserts the moral right to be identified as the author of this work.

All rights reserved. Apart from any fair dealing for the purposes of research or private study or critism or review, no part of this publication may be reproduced, distributed, or transmitted in any form or by any means, including photocopying, recording, or other electronic or mechanical methods, or by any information storage and retrieval system without the prior written permission of the publisher.

Table of Contents

Introduction 1
A Generalised Energy Density Equation 2
Coupling 3
A Generalised Magnetic Field Equation 3
Force and Inertia 6
Newton's law of Gravity 7
Derivation of Coulomb force law 9
Electromagnetic force 9
Modified Newtonian Dynamics (MOND) 11
Conclusion 15
Glossary 16
Bibliography 36
Acknowledgments 57

About the Author 58

Introduction

All fundamental forces of nature including the law of gravity where discovered over three hundred years ago. Many attempts were made to derive Newton's law. However, since all attemps failed, I believe the fundamental forces cannot be derived from the current view-point of physics. Otherwise, Newton's law and Coulombs law would certainly have been derived by now.

Consequently, I propose a more basic and comprehensive viewpoint in Physics- the generalized energy density equation, which has been named by mainstream media as the "holy grail of modern Physics". The existence of particles, fields and quantum effects in the universe may be derived from this underlying principle. It can also be used for practical applications, for example: extracting energy from the quantum vacuum, Casimir batteries and engines, single heat baths and lastly in space travel and gravitational shielding.

If a theory that is used to do calculations for a rocket launch can be derived from one underlying principle- the generalized energy density, imagine what that principle could do for the entire universe when put to good use by NASA, SpaceX and Blue Origin.

This physics letter presents a definitive, accessible proof to how fundamental forces can be derived from the generalized energy density, solving most of the major shortcomings of general relativity and the standard model. This single theory also provides a consistent explanation of the force of gravity and inertia, something that still eludes those who religiously follow the standard model.

A Generalised Energy Density Equation

The easiest way to derive the energy stored in the electric field will assume that the energy density ρ is proportional to the square of the force F but inversely proportional to the coupling α :

$$\rho = \frac{F^2}{8\pi\alpha\hbar c} \qquad (1)$$

Where $\hbar$ is the reduced Planck constant and c is the speed of light.

The electric force on a particle of charge e in the electric field E is

$$F = Ee$$

The coupling constant that determines the strength of the electromagnetic force is:

$$\alpha = \frac{e^2}{4\pi\varepsilon_0\hbar c}$$

Putting all of these into consideration we have:

$$\rho = \frac{\varepsilon_0 E^2}{2}$$

This is the usual formula for the energy density stored in the electromagnetic field. Therefore eqn1 becomes a general rule for all fields whether electromagnetic, gravitational etc. It is the basis for our study of the origin of the fundamental forces of nature.

From eqn1, force is generalised:

$$F = \sqrt{4\pi\alpha\hbar c\rho}$$

Coupling

From the above equation, we want to derive Newton's second law of motion and other forces but before we do, let us define the coupling.

Let the coupling be the number of the degrees of freedom:

$$\alpha = \frac{B_1}{B_o}$$

Where, B is the magnetic field from quantum vacuum fluctuations.

Where does the magnetic field come from?

A Generalised Magnetic Field Equation

Let us now derive the magnetic field from classical statistical mechanics.

The statistical equipartition theorem defines the temperature T of a system in terms of its energy W such that:

$$W = 2\pi K_B T$$

Where K_B is the Boltzmann constant. This is the equivalent energy for mass m according to:

$$W = mc^2$$

The effective temperature experienced due to a uniform acceleration in a vacuum field according to the Unruh effect is:

$$T = \frac{\hbar a}{2\pi v_f K_B}$$

Where a is that acceleration, which for a mass m would be attributed to a force F according to Casimir effect due to quantum vacuum fluctuations.

The velocity of a particle through a quantum vacuum is related to the electric E and magnetic fields B created by vacuum fluctuations by:

$$v_f = \frac{E}{B}$$

When substituted in the equation for T we find the magnetic field:

$$B = \frac{2\pi E K_B T}{\hbar a}$$

The electric field on a charge at a point r in space is:

$$E = \frac{e}{4\pi \varepsilon_o r^2}$$

This leads to the general equation of the magnetic field

$$B_o = \frac{emc^2}{4\pi\varepsilon_o r^2 \hbar a} \qquad (2)$$

Force and Inertia

Assume an electron at a distance r from a point field moves in circular motion at the speed of light forms a boundary, such that

$$a = \frac{c^2}{r} \qquad \text{when} \qquad r = \frac{\hbar}{mc}$$

When substituted in eqn2 we have:

$$B_1 = \frac{em^2 c}{4\pi\varepsilon_o \hbar^2} \qquad (3)$$

Taking all of the above into consideration, the coupling is:

$$\alpha = \frac{B_1}{B_o} = \frac{mar^2}{\hbar c}$$

Suppose our boundary forms a closed surface. One can think about the boundary as a storage device for information. The energy density or pressure within an enclosed boundary is:

$$\rho = \frac{F}{A}$$

Taking the boundary to be a sphere of radius r, the surface area would be given by:

$$A = 4\pi r^2$$

From algebraic substitution of these into the above relations, one derives Newton's second law of motion:

$$F = \sqrt{4\pi\alpha\hbar c\rho} = \sqrt{4\pi\frac{mar^2}{\hbar c}\hbar c\frac{F}{4\pi r^2}}$$

$$F = ma$$

Newton's law of Gravity

For the case of a Schwarzischild radius Black hole,

$$r = \frac{GM}{c^2} \quad \text{when} \quad a = \frac{c^2}{r}$$

Equation2 becomes:

$$B_2 = \frac{emc^2}{4\pi\varepsilon_o GM\hbar} \tag{4}$$

Taking all of the above into consideration, the coupling is:

$$\alpha = \frac{B_1}{B_2} = \frac{GMm}{\hbar c}$$

Suppose our boundary forms a closed surface. One can think about the boundary as a storage device for information. The energy density or pressure within an enclosed boundary is:

$$\rho = \frac{F}{A}$$

Taking the boundary to be a sphere of radius r, the surface area would be given by:
$$A = 4\pi r^2$$

From algebraic substitution of these into the above relations, one derives Newton's law of gravitation:

$$F = \sqrt{4\pi\alpha\hbar c\rho} = \sqrt{4\pi\frac{GMm}{\hbar c}\hbar c\frac{F}{4\pi r^2}}$$

$$F = \frac{GMm}{r^2}$$

Derivation of Coulomb force law

Electromagnetic force

For the case of a Schwarzischild radius electron Black hole,

$$r = \frac{e^2}{4\pi\varepsilon_o mc^2} \quad \text{when} \quad a = \frac{c^2}{r}$$

Equation2 becomes the Schwinger limit:

$$B_3 = \frac{m^2 c^2}{\hbar e} \quad (5)$$

Taking all of the above into consideration, the coupling is:

$$\alpha = \frac{B_1}{B_3} = \frac{e^2}{4\pi\varepsilon_o \hbar c}$$

Suppose our boundary forms a closed surface. One can think about the boundary as a storage device for information. The energy density or pressure within an enclosed boundary is:

$$\rho = \frac{F}{A}$$

Taking the boundary to be a sphere of radius r, the surface area would be given by:
$$A = 4\pi r^2$$

From algebraic substitution of these into the above relations, one derives Coulombs law of electricity:

$$F = \sqrt{4\pi\alpha\hbar c\rho} = \sqrt{4\pi \frac{e^2}{4\pi\varepsilon_o \hbar c}\hbar c \frac{F}{4\pi r^2}}$$

$$F = \frac{e^2}{4\pi\varepsilon_o r^2}$$

Modified Newtonian Dynamics (MOND)

In the disc galaxies most of the mass is at the centre of the galaxy, this means that if you want to calculate how a star moves far away from the centre it is a good approximation to only ask what is the gravitational pull that comes from the centre bulge of the galaxy.

Einstein taught us that gravity is really due to the curvature of space and time but in many cases it is still quantitatively incorrect to describe gravity as a force, this is known as the Newtonian limit and is a good approximation as long as the pull of gravity is weak and objects move much slower than the speed of light. It is a bad approximation for example close by the horizon of a black hole but it is a good approximation for the dynamics of galaxies that we are looking at here. It is then not difficult to calculate the stable orbit of a star far away from the centre of a disc galaxy.

For a star to remain on its orbit, the gravitational pull must be balanced by the centrifugal force,

$$\frac{mv^2}{r} = \frac{GMm}{r^2}.$$

You can solve this equation for the velocity of the star and this will give you the velocity that is necessary for a star to remain on a stable orbit,

$$v = \sqrt{\frac{GM}{r}}.$$

As you can see the velocity drops inversely with the square root of the distance to the centre. But this is not what we observe, what we observe instead is that the velocity continue to increase with distance from the galactic centre and then they become constant.

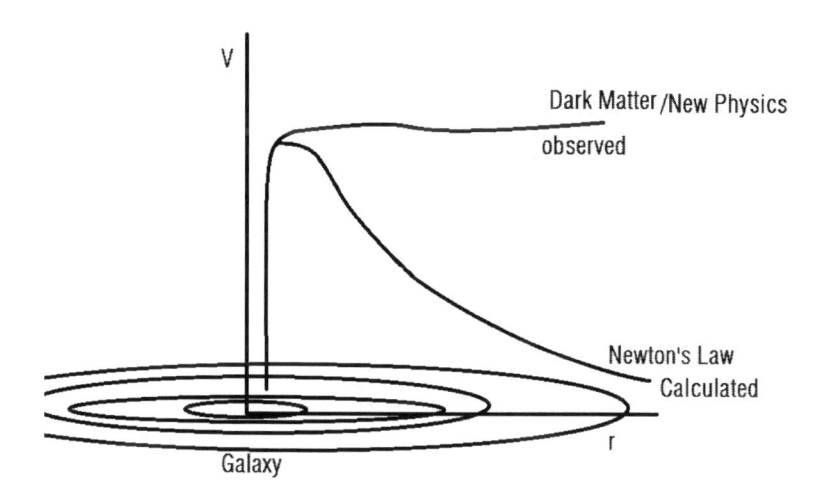

[Figure: Galaxy rotation curve showing V vs r. "Dark Matter/New Physics observed" curve remains flat; "Newton's Law Calculated" curve decreases. Labeled "Galaxy" below.]

This is known as the flat rotation curve. This is not only the case for our own galaxy but it is the case for hundred of galaxies that have been observed. The curves don't always become perfectly constant sometimes they have rigorous lines but it is abundantly clear that these observations cannot be explained by the normal matter only.

Dark matter solves this problem by postulating that there is additional mass in galaxies distributed in a spherical halo. This has the effect of speeding up the stars because the gravitational pull is now stronger due to the mass from the dark matter halo. There is always a distribution of dark matter that will reproduce whatever velocity curve we observe.

This invisible and undetected matter removes any need to modify Newton's and Einstein's gravitational theories. Invoking dark matter is a less radical, less scary alternative for most physicists than inventing a new theory of gravity.

If dark matter is not detected and does not exist, then Einstein's and Newton's gravity theories must be modified. Can this be done successfully? Yes! Our generalized force equation (1) can explain the astrophysical, astronomical and cosmological data without dark matter as given below.

In contrast to Dark matter, the generalized force (1) postulates that gravity works differently.

From equation (1) the gravitational force is the inverse of the distance

$$F = \frac{1}{r}\sqrt{\alpha \hbar c m a_o}$$

While normally it is the inverse of the square of the distance

$$F = \frac{GMm}{r^2}$$

If you put this modified gravitational force into the force balance equation as before

$$\frac{mv^2}{r} = \frac{1}{r}\sqrt{\alpha \hbar c m a_o}$$

you will see that the dependence on the distance cancels out and the velocity just becomes constant. Finally one inserts

$$\alpha = \frac{GMm}{\hbar c}$$

and one obtains the familiar relation

$$v^4 = a_o GM$$

We have recovered the Tully-Fisher relation, practically from first principles!

This generalised force is an alternative to the hypothesis of dark matter in terms of explaining why galaxies do not appear to obey the currently understood laws of physics.

Now of course you cannot just go and throw out the normal

$$\frac{1}{r^2}$$

gravitational force law because we know that it works on the solar system. Therefore the normal $\frac{1}{r^2}$ law crosses over into a $\frac{1}{r}$ law. This crossover happens not at a certain distance but it happens at certain acceleration. The New force law comes into play at low acceleration $a_o = 1.2 \times 10^{-10} ms^{-2}$, this acceleration where the crossover happens is a free parameter in MOND. You can determine the value of this pararmeter by just trying out which fits the data best. It turns out that the best fit value is closely related to the cosmological constant

$$a_o \approx \sqrt{\frac{\Lambda}{3}}$$

We anticipate that the generalised force will modify how stars collapse and the nature of black holes.

Conclusion

The key statement is simply that we need to have a temperature in order to have a magnetic field. Since we want to understand the origin of the fundamental forces of nature, we need to know where the magnetic field and the coupling come from.

Note that this derivation assumes that the number of degrees of freedom is equal to the coupling. Finally, the universal force equation that units all interactions is given by:

$$F = \sqrt{4\pi \frac{B_1}{B_o} \hbar c \rho}$$

The above equation has practical applications, for example: in extracting energy from the quantum vacuum, Casimir batteries and engines, single heat baths and lastly in space travel and gravitational shielding which will prove to be important for NASA and SpaceX technologies.

Coming soon in the second Edition:
1. Practical Applications to Space Travel

Glossary

Absolute space and time—the Newtonian concepts of space and time, in which space is independent of the material bodies within it, and time flows at the same rate throughout the universe without regard to the locations of different observers and their experience of "now."

Acceleration—the rate at which the speed or velocity of a body changes.

Accelerating universe—the discovery in 1998, through data from very distant supernovae, that the expansion of the universe in the wake of the big bang is not slowing down, but is actually speeding up at this point in its history; groups of astronomers in California and Australia independently discovered that the light from the supernovae appears dimmer than would be expected if the universe were slowing down.

Action—the mathematical expression used to describe a physical system by requiring only the knowledge of the initial and final states of the system; the values of the physical variables at all intermediate states are determined by minimizing the action.

Anthropic principle—the idea that our existence in the universe imposes constraints on its properties; an extreme version claims that we owe our existence to this principle.

Asymptotic freedom (or safety)—a property of quantum field theory in which the strength of the coupling between elementary particles vanishes with increasing energy and/or decreasing distance, such that the elementary particles approach free particles with no external forces acting on them; moreover for decreasing energy and/or increasing distance between the particles, the strength of the particle force increases indefinitely.

Baryon—a subatomic particle composed of three quarks, such as the proton and neutron.

Big bang theory—the theory that the universe began with a violent explosion of spacetime, and that matter and energy originated from an infinitely small and dense point.

Big crunch—similar to the big bang, this idea postulates an end to the universe in a singularity.

Binary stars—a common astrophysical system in which two stars rotate around each other; also called a "double star."

Blackbody—a physical system that absorbs all radiation that hits it, and emits characteristic radiation energy depending upon temperature; the concept of blackbodies is useful, among other things, in learning the temperature of stars.

Black hole—created when a dying star collapses to a singular point, concealed by an "event horizon;" the black hole is so dense and has such strong gravity that nothing, including light, can escape it; black holes are predicted by general relativity, and though they cannot be "seen," several have been inferred from astronomical observations of binary stars and massive collapsed stars at the centers of galaxies.

Boson—a particle with integer spin, such as photons, mesons, and gravitons, which carries the forces between fermions.

Brane—shortened from "membrane," a higher-dimensional extension of a onedimensional string.

Cassini spacecraft—NASA mission to Saturn, launched in 1997, that in addition to making detailed studies of Saturn and its moons, determined a bound on the variations of Newton's gravitational constant with time.

Causality—the concept that every event has in its past events that caused it, but no event can play a role in causing events in its past.

Classical theory—a physical theory, such as Newton's gravity theory or Einstein's general relativity, that is concerned with the macroscopic universe, as opposed to theories concerning events at the submicroscopic level such as quantum mechanics and the standard model of particle physics.

Copernican revolution—the paradigm shift begun by Nicolaus Copernicus in the early sixteenth century, when he identified the sun, rather than the Earth, as the center of the known universe.

Cosmic microwave background (CMB)—the first significant evidence for the big bang theory; initially found in 1964 and studied further by NASA teams in 1989 and the early 2000s, the CMB is a smooth signature of microwaves everywhere in the sky, representing the "afterglow" of the big bang: Infrared light produced about 400,000 years after the big bang had redshifted through the stretching of spacetime during fourteen billion years of expansion to the microwave part of the electromagnetic spectrum, revealing a great deal of information about the early universe.

Cosmological constant—a mathematical term that Einstein inserted into his gravity field equations in 1917 to keep the universe static and eternal; although he later regretted this and called it his "biggest blunder," cosmologists today still use the
cosmological constant, and some equate it with the mysterious dark energy.

Coupling constant—a term that indicates the strength of an interaction between particles or fields; electric charge and Newton's gravitational constant are coupling constants.

Crystalline spheres—concentric transparent spheres in ancient Greek cosmology that held the moon, sun, planets, and stars in place and made them revolve around the Earth; they were part of the western conception of the universe until the Renaissance.

Curvature—the deviation from a Euclidean spacetime due to the warping of the geometry by massive bodies.

Dark energy—a mysterious form of energy that has been associated with negative pressure vacuum energy and Einstein's cosmological constant; it is hypothesized to explain the data on the accelerating expansion of the universe; according to the standard model, the dark energy, which is spread uniformly
throughout the universe, makes up about 70 percent of the total mass and energy content of the universe.

Dark matter—invisible, not-yet-detected, unknown particles of matter, representing about 30 percent of the total mass of matter according to the standard model; its presence is necessary if Newton's and Einstein's gravity theories are to fit data from galaxies, clusters of galaxies, and cosmology; together, dark
matter and dark energy mean that 96 percent of the matter and energy in the universe is invisible.

Deferent—in the ancient Ptolemaic concept of the universe, a large circle representing the orbit of a planet around the Earth.

Doppler principle or **Doppler effect**—the discovery by the nineteenth-century Austrian scientist Christian Doppler that when sound or light waves are moving toward an observer, the apparent frequency of the waves will be shortened, while if they are moving away from an observer, they will be lengthened; in astronomy this means that the light emitted by galaxies moving away from us is redshifted, and that from nearby galaxies moving toward us is blueshifted.

Dwarf galaxy—a small galaxy (containing several billion stars) orbiting a larger galaxy; the Milky Way has over a dozen dwarf galaxies as companions, including the Large Magellanic Cloud and Small Magellanic Cloud.

Dynamics—the physics of matter in motion.

Electromagnetism—the unified force of electricity and magnetism, discovered to be the same phenomenon by Michael Faraday and James Clerk Maxwell in the nineteenth century.

Electromagnetic radiation—a term for wave motion of electromagnetic fields which propagate with the speed of light—300,000 kilometers per second—and differ only in wavelength; this includes visible light, ultraviolet light, infrared radiation, X-rays, gamma rays, and radio waves.

Electron—an elementary particle carrying negative charge that orbits the nucleus of an atom.

Eötvös experiments—torsion balance experiments performed by Hungarian Count Roland von Eötvös in the late nineteenth and early twentieth centuries that showed that inertial and gravitational mass were the same to one part in 1011; this was a more accurate determination of the equivalence principle than results achieved by Isaac Newton and, later, Friedrich Wilhelm Bessel.

Epicycle—in the Ptolemaic universe, a pattern of small circles traced out by a planet at the edge of its "deferent" as it orbited the Earth; this was how the Greeks accounted for the apparent retrograde motions of the planets.

Equivalence principle—the phenomenon first noted by Galileo that bodies falling in a gravitational field fall at the same rate, independent of their weight and composition; Einstein extended the principle to show that gravitation is identical (equivalent) to acceleration.

Escape velocity—the speed at which a body must travel in order to escape a strong gravitational field; rockets fired into orbits around the Earth have calculated escape velocities, as do galaxies at the periphery of galaxy clusters.

Ether (or aether)—a substance whose origins were in the Greek concept of "quintessence," the ether was the medium through which energy and matter moved, something more than a vacuum and less than air; in the late nineteenth century the Michelson-Morley experiment disproved the existence of the ether.

Euclidean geometry—plane geometry developed by the third-century bc Greek mathematician Euclid; in this geometry, parallel lines never meet.

Fermion—a particle with half-integer spin, like protons and electrons, that make up matter.

Field—a physical term describing the forces between massive bodies in gravity and electric charges in electromagnetism; Michael Faraday discovered the concept of field when studying magnetic conductors.

Field equations—differential equations describing the physical properties of interacting massive particles in gravity and electric charges in electromagnetism;

Maxwell's equations for electromagnetism and Einstein's equations of gravity are prominent examples in physics.

Fifth force or **"skew" force**—a new force in MOG that has the effect of modifying gravity over limited length scales; it is carried by a particle with mass called the phion.

Fine-tuning—the unnatural cancellation of two or more large numbers involving an absurd number of decimal places, when one is attempting to explain a physical phenomenon; this signals that a true understanding of the physical phenomenon has not been achieved.

Fixed stars—an ancient Greek concept in which all the stars were static in the sky, and moved around the Earth on a distant crystalline sphere.

Frame of reference—the three spatial coordinates and one time coordinate that an observer uses to denote the position of a particle in space and time.

Galaxy—organized group of hundreds of billions of stars, such as our Milky Way.

Galaxy cluster—many galaxies held together by mutual gravity but not in as organized a fashion as stars within a single galaxy.

Galaxy rotation curve—a plot of the Doppler shift data recording the observed velocities of stars in galaxies; those stars at the periphery of giant spiral galaxies are observed to be going faster than they "should be" according to Newton's and Einstein's gravity theories.

General relativity—Einstein's revolutionary gravity theory, created in 1916 from a mathematical generalization of his theory of special relativity; it changed our concept of gravity from Newton's universal force to the warping of the geometry of spacetime in the presence of matter and energy.

Geodesic—the shortest path between two neighboring points, which is a straight line in Euclidian geometry, and a unique curved path in four-dimensional spacetime.

Globular cluster—a relatively small, dense system of up to millions of stars occurring commonly in galaxies.

Gravitational lensing—the bending of light by the curvature of spacetime; galaxies and clusters of galaxies act as lenses, distorting the images of distant bright galaxies or quasars as the light passes through or near them.

Gravitational mass—the active mass of a body that produces a gravitational force on other bodies.

Gravitational waves—ripples in the curvature of spacetime predicted by general relativity; although any accelerating body can produce gravitational radiation or waves, those that could be detected by experiments would be caused by cataclysmic cosmic events.

Graviton—the hypothetical smallest packet of gravitational energy, comparable to the photon for electromagnetic energy; the graviton has not yet been seen experimentally.

Group (in mathematics)—in abstract algebra, a set that obeys a binary operation that satisfies certain axioms; for example, the property of addition of integers makes a group; the branch of mathematics that studies groups is called group theory.

Hadron—a generic word for fermion particles that undergo strong nuclear interactions.

Hamiltonian—an alternative way of deriving the differential equations of motion for a physical system using the calculus of variations; Hamilton's principle is also called the "principle of stationary action" and was originally formulated by Sir William Rowan Hamilton for classical mechanics; the principle applies to classical fields such as the gravitational and electromagnetic fields, and has had important applications in quantum mechanics and quantum field theory.

Homogeneous—in cosmology, when the universe appears the same to all observers, no matter where they are in the universe.

Inertia—the tendency of a body to remain in uniform motion once it is moving, and to stay at rest if it is at rest; Galileo discovered the law of inertia in the early seventeenth century.

Inertial mass—the mass of a body that resists an external force; since Newton, it has been known experimentally that inertial and gravitational mass are equal; Einstein used this equivalence of inertial and gravitational mass to postulate his equivalence principle, which was a cornerstone of his gravity theory.

Inflation theory—a theory proposed by Alan Guth and others to resolve the flatness, horizon, and homogeneity problems in the standard big bang model; the very early universe is pictured as expanding exponentially fast in a fraction of a second.

Interferometry—the use of two or more telescopes, which in combination create a receiver in effect as large as the distance between them; radio astronomy in particular makes use of interferometry.

Inverse square law—discovered by Newton, based on earlier work by Kepler, this law states that the force of gravity between two massive bodies or point particles decreases as the inverse square of the distance between them.

Isotropic—in cosmology, when the universe looks the same to an observer, no matter in which direction she looks.

Kelvin temperature scale—designed by Lord Kelvin (William Thomson) in the mid-1800s to measure very cold temperatures, its starting point is absolute zero, the coldest possible temperature in the universe, corresponding to −273.15 degrees Celsius; water's freezing point is 273.15K (0°C), while its boiling point is 373.15K (100°C).

Lagrange points—discovered by the Italian-French mathematician Joseph-Louis Lagrange, these five special points are in the vicinity of two orbiting masses where a third, smaller mass can orbit at a fixed distance from the larger masses; at the Lagrange points, the gravitational pull of the two large masses precisely equals the centripetal force required to keep the third body, such as a satellite, in a bound orbit; three of the Lagrange points are unstable, two are stable.

Lagrangian—named after Joseph-Louis Lagrange, and denoted by L, this mathematical expression summarizes the dynamical properties of a physical system; it is defined in classical mechanics as the kinetic energy T minus the potential energy V; the equations of motion of a system of particles may be derived from the Euler-Lagrange equations, a family of partial differential equations.

Light cone—a mathematical means of expressing past, present, and future space and time in terms of spacetime geometry; in four-dimensional Minkowski spacetime, the light rays emanating from or arriving at an event separate spacetime into a past cone and a future cone which meet at a point corresponding
to the event.

Lorentz transformations—mathematical transformations from one inertial frame of reference to another such that the laws of physics remain the same; named after Hendrik Lorentz, who developed them in 1904, these transformations form the basic mathematical equations underlying special relativity.

Mercury anomaly—a phenomenon in which the perihelion of Mercury's orbit advances more rapidly than predicted by Newton's equations of gravity; when Einstein showed that his gravity theory predicted the anomalous precession, it was the first empirical evidence that general relativity might be correct.

Meson—a short-lived boson composed of a quark and an antiquark, believed to bind protons and neutrons together in the atomic nucleus.

Metric tensor—mathematical symmetric tensor coefficients that determine the infinitesimal distance between two points in spacetime; in effect the metric tensor distinguishes between Euclidean and non-Euclidean geometry.

Michelson-Morley experiment—1887 experiment by Albert Michelson and Edward Morley that proved that the ether did not exist; beams of light traveling in the same direction, and in the perpendicular direction, as the supposed ether showed no difference in speed or arrival time at their destination.

Milky Way—the spiral galaxy that contains our solar system.

Minkowski spacetime—the geometrically flat spacetime, with no gravitational effects, first described by the Swiss mathematician Hermann Minkowski; it became the setting of Einstein's theory of gravity.

MOG—my relativistic modified theory of gravitation, which generalizes Einstein's general relativity; MOG stands for "Modified Gravity."

MOND—a modification of Newtonian gravity published by Mordehai Milgrom in 1983; this is a nonrelativistic phenomenological model used to describe rotational velocity curves of galaxies; MOND stands for "Modified Newtonian Dynamics."

Neutrino—an elementary particle with zero electric charge; very difficult to detect, it is created in radioactive decays and is able to pass through matter almost undisturbed; it is considered to have a tiny mass that has not yet been accurately measured.

Neutron—an elementary and electrically neutral particle found in the atomic nucleus, and having about the same mass as the proton.

Nuclear force—another name for the strong force that binds protons and neutrons together in the atomic nucleus.

Nucleon—a generic name for a proton or neutron within the atomic nucleus.

Neutron star—the collapsed core of a star that remains after a supernova explosion; it is extremely dense, relatively small, and composed of neutrons.

Newton's gravitational constant—the constant of proportionality, G, which occurs in the Newtonian law of gravitation, and says that the attractive force between two bodies is proportional to the product of their masses and inversely proportional to the square of the distance between them; its numerical value is: $G = 6.67428 \pm 0.00067 \times 10^{-11}$ m3 kg−1 s−2.

Nonsymmetric field theory (Einstein)—a mathematical description of the geometry of spacetime based on a metric tensor that has both a symmetric part and an antisymmetric part; Einstein used this geometry to formulate a unified field
theory of gravitation and electromagnetism.

Nonsymmetric Gravitation Theory (NGT)—my generalization of Einstein's purely gravitation theory (general relativity) that introduces the antisymmetric field as an extra component of the gravitational field; mathematically speaking, the nonsymmetric field structure is described by a non-Riemannian geometry.

Parallax—the apparent movement of a nearer object relative to a distant background when one views the object from two different positions; used with triangulation for measuring distances in astronomy.

Paradigm shift—a revolutionary change in belief, popularized by the philosopher Thomas Kuhn, in which the majority of scientists in a given field discard a traditional theory of nature in favor of a new one that passes the tests of experiment and observation; Darwin's theory of natural selection, Newton's gravity theory, and Einstein's general relativity all represented paradigm shifts.

Parsec—a unit of astronomical length equal to 3.262 light years.

Particle-wave duality—the fact that light in all parts of the electromagnetic spectrum (including radio waves, X-rays, etc., as well as visible light) sometimes acts like waves and sometimes acts like particles or photons; gravitation may be similar, manifesting as waves in spacetime or graviton particles.

Perihelion—the position in a planet's elliptical orbit when it is closest to the sun.

Perihelion advance—the movement, or changes, in the position of a planet's perihelion in successive revolutions of its orbit over time; the most dramatic perihelion advance is Mercury's, whose orbit traces a rosette pattern.

Perturbation theory—a mathematical method for finding an approximate solution to an equation that cannot be solved exactly, by expanding the solution in a series in which each successive term is smaller than the preceding one.

Phion—name given to the massive vector field in MOG; it is represented both by a boson particle, which carries the fifth force, and a field.

Photoelectric effect—the ejection of electrons from a metal by X-rays, which proved the existence of photons; Einstein's explanation of this effect in 1905 won him the Nobel Prize in 1921; separate experiments proving and demonstrating
the existence of photons were performed in 1922 by Thomas Millikan and Arthur Compton, who received the Nobel Prize for this work in 1923 and 1927, respectively.

Photon—the quantum particle that carries the energy of electromagnetic waves; the spin of the photon is 1 times Planck's constant h.

Pioneer 10 and 11 spacecraft—launched by NASA in the early 1970s to explore the outer solar system, these spacecraft show a small, anomalous acceleration as they leave the inner solar system.

Planck's constant (h)—a fundamental constant that plays a crucial role in quantum mechanics, determining the size of quantum packages of energy such as the photon; it is named after Max Planck, a founder of quantum mechanics

Principle of general covariance—Einstein's principle that the laws of physics remain the same whatever the frame of reference an observer uses to measure physical quantities.

Principle of least action—more accurately the principle of *stationary* action, this variational principle, when applied to a mechanical system or a field theory, can be used to derive the equations of motion of the system; the credit for discovering the principle is given to Pierre-Louis Moreau Maupertius but it may have been discovered independently by Leonhard Euler or Gottfried Leibniz.

Proton—an elementary particle that carries positive electrical charge and is the nucleus of a hydrogen atom.

Ptolemaic model of the universe—the predominant theory of the universe until the Renaissance, in which the Earth was the heavy center of the universe and all other heavenly bodies, including the moon, sun, planets, and stars, orbited around it; named for Claudius Ptolemy.

Quantize—to apply the principles of quantum mechanics to the behavior of matter and energy (such as the electromagnetic or gravitational field energy); breaking down a field into its smallest units or packets of energy.

Quantum field theory—the modern relativistic version of quantum mechanics used to describe the physics of elementary particles; it can also be used in nonrelativistic fieldlike systems in condensed matter physics.

Quantum gravity—attempts to unify gravity with quantum mechanics.

Quantum mechanics—the theory of the interaction between quanta (radiation) and matter; the effects of quantum mechanics become observable at the submicroscopic distance scales of atomic and particle physics, but macroscopic quantum effects can be seen in the phenomenon of quantum entanglement.

Quantum spin—the intrinsic quantum angular momentum of an elementary particle; this is in contrast to the classical orbital angular momentum of a body rotating about a point in space.

Quark—the fundamental constituent of all particles that interact through the strong nuclear force; quarks are fractionally charged, and come in several varieties; because they are confined within particles such as protons and neutrons, they cannot be detected as free particles.

Quasars—"quasi-stellar objects," the farthest distant objects that can be detected with radio and optical telescopes; they are exceedingly bright, and are believed to be young, newly forming galaxies; it was the discovery of quasars in 1960 that quashed the steady-state theory of the universe in favor of the big bang.

Quintessence—a fifth element in the ancient Greek worldview, along with earth, water, fire and air, whose purpose was to move the crystalline spheres that supported the heavenly bodies orbiting the Earth; eventually this concept became known as the "ether," which provided the *something* that bodies needed to be in contact with in order to move; although Einstein's special theory of relativity dispensed with the ether, recent explanations of the acceleration of the universe call the varying negative pressure vacuum energy "quintessence."

Redshift—a useful phenomenon based on the Doppler principle that can indicate whether and how fast bodies in the universe are receding from an observer's position on Earth; as galaxies move rapidly away from us, the frequency of the wavelength of their light is shifted toward the red end of the electromagnetic

spectrum; the amount of this shifting indicates the distance of the galaxy.

Riemann curvature tensor—a mathematical term that specifies the curvature of four-dimensional spacetime.

Riemannian geometry—a non-Euclidean geometry developed in the mid-nineteenth century by the German mathematician George Bernhard Riemann that describes curved surfaces on which parallel lines *can* converge, diverge, and even intersect, unlike Euclidean geometry; Einstein made Riemannian geometry the mathematical formalism of general relativity.

Satellite galaxy—a galaxy that orbits a host galaxy or even a cluster of galaxies.

Scalar field—a physical term that associates a value without direction to every point in space, such as temperature, density, and pressure; this is in contrast to a vector field, which has a direction in space; in Newtonian physics or in electrostatics, the potential energy is a scalar field and its gradient is the vector force field; in quantum field theory, a scalar field describes a boson particle with spin zero.

Scale invariance—distribution of objects or patterns such that the same shapes and distributions remain if one increases or decreases the size of the length scales or

space in which the objects are observed; a common example of scale invariance
is fractal patterns.

Schwarzschild solution—an exact spherically symmetric static solution of Einstein's field equations in general relativity, worked out by the astronomer Karl Schwarzschild in 1916, which predicted the existence of black holes.

Self-gravitating system—a group of objects or astrophysical bodies held together by mutual gravitation, such as a cluster of galaxies; this is in contrast to a "bound system" like our solar system, in which bodies are mainly attracted to and revolve around a central mass.

Singularity—a place where the solutions of differential equations break down; a spacetime singularity is a position in space where quantities used to determine the gravitational field become infinite; such quantities include the curvature of spacetime and the density of matter.

Spacetime—in relativity theory, a combination of the three dimensions of space with time into a four-dimensional geometry; first introduced into relativity by Hermann Minkowski in 1908.

Special theory of relativity—Einstein's initial theory of relativity, published in 1905, in which he explored the "special" case of transforming the laws of physics from one uniformly moving frame of reference to another; the equations
of special relativity revealed that the speed of light is a constant, that objects appear contracted in the direction of motion when moving at close to the speed of light, and that $E = mc2$, or energy is equal to mass times the speed of light squared.

Spin—see quantum spin.

String theory—a theory based on the idea that the smallest units of matter are not point particles but vibrating strings; a popular research pursuit in physics for two decades, string theory has some attractive mathematical features, but has yet to make a testable prediction.

Strong force—see nuclear force.

Supernova—spectacular, brilliant death of a star by explosion and the release of heavy elements into space; supernovae type 1a are assumed to have the same intrinsic brightness and are therefore used as standard candles in estimating cosmic distances.

Supersymmetry—a theory developed in the 1970s which, proponents claim, describes the most fundamental spacetime symmetry of particle physics: For every boson particle there is a supersymmetric fermion partner, and for every fermion there exists a supersymmetric boson partner; to date, no supersymmetric particle partner has been detected.

Tully-Fisher law—a relation stating that the asymptotically flat rotational velocity of a star in a galaxy, raised to the fourth power, is proportional to the mass or luminosity of the galaxy.

Unified theory (or unified field theory)—a theory that unites the forces of nature; in Einstein's day those forces consisted of electromagnetism and gravity; today the weak and strong nuclear forces must also be taken into account, and perhaps someday MOG's fifth force or skew force will be included; no one has yet discovered a successful unified theory.

Vacuum—in quantum mechanics, the lowest energy state, which corresponds to the vacuum state of particle physics; the vacuum in modern quantum field theory is the state of perfect balance of creation and annihilation of particles and antiparticles.

Variable Speed of Light (VSL) cosmology—an alternative to inflation theory, in which the speed of light was much faster at the beginning of the universe than it is today; like inflation, this theory solves the horizon and flatness problems in the very early universe in the standard big bang model.

Vector field—a physical value that assigns a field with the position and direction of a vector in space; it describes the force field of gravity or the electric and magnetic force fields in James Clerk Maxwell's field equations.

Virial theorem—a means of estimating the average speed of galaxies within galaxy clusters from their estimated average kinetic and potential energies.

Vulcan—a hypothetical planet predicted by the nineteenth-century astronomer Urbain Jean Joseph Le Verrier to be the closest orbiting planet to the sun; the presence of Vulcan would explain the anomalous precession of the perihelion of Mercury's orbit; Einstein later explained the anomalous precession in general relativity by gravity alone.

Weak force—one of the four fundamental forces of nature, associated with radioactivity such as beta decay in subatomic physics; it is much weaker than the strong nuclear force but still much stronger than gravity.

X-ray clusters—galaxy clusters with large amounts of extremely hot gas within them that emit X-rays; in such clusters, this hot gas represents at least twice the mass of the luminous stars.

Bibliography

Misner, C.W., Thorne, K.S., and Wheeler, J.A. (1973) *Gravitation*, Freeman, San Francisco, p. 5.

Wilson, H.A. (1921) An electromagnetic theory of gravitation, *Phys. Rev.* **17**, 54-59.

Dicke, R.H. (1957) Gravitation without a principle of equivalence, *Rev. Mod. Phys.* **29**, 363-376. See also Dicke, R.H. (1961) Mach's principle and equivalence, in C. Møller (ed.), *Proc. of the Intern'l School of Physics "Enrico Fermi" Course XX, Evidence for Gravitational Theories*, Academic Press, New York, pp.1-49.

Lightman, A.P., and Lee, D.L. (1973) Restricted proof that the weak equivalence principle implies the Einstein equivalence principle, *Phys. Rev. D* **8**, 364-376.

Will, C.M. (1974) Gravitational red-shift measurements as tests of nonmetric theories of gravity, *Phys. Rev. D* **10**, 2330-2337.

Haugan, M.P., and Will, C.M. (1977) Princip les of equivalence, Eötvös experiments, and gravitational red-shift experiments: The free fall of electromagnetic systems to post-post-Coulombian order, *Phys. Rev. D* **15**, 2711-2720.

Volkov, A.M., Izmest'ev, A.A., and Skrotskii, G.V. (1971) The propagation of electromagnetic waves in a Riemannian space, *Sov. Phys. JETP* **32**, 686-689.

Heitler, W. (1954) *The Quantum Theory of Radiation*, 3rd ed., Oxford University Press, London, p. 113.

Alpher, R.A. (Jan.-Feb. 1973) Large numbers, cosmology, and Gamow, *Am. Sci.* **61**, 52-58.

Harrison, E.R. (Dec. 1972) The cosmic numbers, *Phys. Today* **25**, 30-34.

Webb, J.K., Flambaum, V.V., Churchill, C.W., Drinkwater, M.J., and Barrow, J.D. (1999) Search for time variation of the fine structure constant, *Phys. Rev. Lett.* **82**, 884-887.

Brault, J.W. (1963) Gravitational red shift of solar lines, *Bull. Amer. Phys. Soc.* **8**, 28.

Pound, R.V., and Rebka, G.A. (1960) Apparent weight of photons, *Phys. Rev. Lett.* **4**, 337-341.

Pound, R.V., and Snider, J.L. (1965) Effect of gravity on nuclear resonance, *Phys. Rev. Lett.* **13**, 539-540.

Goldstein, H. (1957) *Classical Mechanics*, Addison-Wesley, Reading MA, pp. 206-207.

Mizobuchi, Y. (1985) New theory of space-time and gravitation - Yilmaz's approach, *Hadronic Jour.* **8**, 193-219.

Alley, C.O. (1995) The Yilmaz theory of gravity and its compatibility with quantum theory, in D.M. Greenberger and A. Zeilinger (eds.), *Fundamental Problems in Quantum Theory: A Conference Held in Honor of Professor John A. Wheeler*, Vol. 755 of the Annals of the New York Academy of Sciences, New York, pp. 464-475.

Schilling, G.(1999) Watching the universe's second biggest bang, *Science* **283**, 2003-2004.

Robertson, S.L. (1999) Bigger bursts from merging neutron stars, *Astrophys. Jour.* **517**, L117-L119.

Hughes, V.W., Robinson, H.G. and Beltran-Lopez, V. (1960) Upper limit for the anisotropy of inertial mass from nuclear resonance experiments, *Phys. Rev. Lett.* **4**, 342-344.

Drever, R.W.P. (1961) A search for anisotropy of inertial mass using a free precession technique, *Phil. Mag.* **6**, 683-687.

Collela, R. Overhauser, A.W., and Werner, S.A. (1975) Observation of gravitationally induced quantum mechanics, *Phys. Rev Lett.* **34**, 1472-1474.

Puthoff, H.E. (1996) SETI, the velocity-of-light limitation, and the Alcubierre warp drive: an integrating overview, *Physics Essays* **9**, 156-158.

Atkinson, R. d'E. (1962) General relativity in Euclidean terms, *Proc. Roy. Soc.* **272**, 60-7

Barton, G.; Scharnhorst, K. (1993). "QED Between Parallel Mirrors: Light Signals Faster Than c, or Amplified by the Vacuum". Journal of Physics A: Mathematical and General. **26** (8): 2037–2046. *Bibcode:1993JPhA...26.2037B*. *doi:10.1088/0305-4470/26/8/024*. *ISSN 0305-4470*.

Beiser, A. (2003). *Concepts of Modern Physics* (6th ed.). Boston: McGraw-Hill. *ISBN 978-0072448481*. *LCCN 2001044743*. *OCLC 48965418*.

Bordag, M; Klimchitskaya, G. L.; Mohideen, U.; Mostepanenko, V. M. (2009). *Advances in the Casimir Effect*. Oxford: Oxford University Press. *ISBN 978-0-19-923874-3*. *LCCN 2009279136*. *OCLC 319209483*.

Boyer, T. H. (1970). "Quantum Zero-Point Energy and Long-Range Forces". Annals of Physics. 56 (2): 474–503.

Bibcode:*1970AnPhy..56..474B*.*doi*:*10.1016/0003-4916(70)90027-8*. ISSN *0003-4916*. OCLC *4648258537*.

Carroll, S. M.; Field, G. B. (1997). *"Is There Evidence for Cosmic Anisotropy in the Polarization of Distant Radio Sources?"* (PDF). Physical Review Letters. 79 (13): 2394–2397.*arXiv*:*astro-ph/9704263*.

Conlon, T. E. (2011). *Thinking About Nothing : Otto Von Guericke and The Magdeburg Experiments on the Vacuum*. San Francisco: Saint Austin Press. ISBN *978-1-4478-3916-3*. OCLC *840927124*.

Davies, P. C. W. (1985). *Superforce: The Search for a Grand Unified Theory of Nature*. New York: Simon and Schuster. ISBN *978-0-671-47685-4*. LCCN *84005473*. OCLC *12397205*.

Dunne, G. V. (2012). "The Heisenberg-Euler Effective Action: 75 years on". International Journal of Modern Physics A. 27 (15): 1260004. *arXiv*:*1202.1557*. *Bibcode*:*2012IJMPA..2760004D*. *doi*:*10.1142/S0217751X12600044*. ISSN *0217-751X*.

Einstein, A. (1995). Klein, Martin J.; Kox, A. J.; Renn, Jürgen; Schulmann, Robert (eds.). *The Collected Papers of Albert Einstein Vol. 4 The Swiss Years: Writings, 1912–1914*. Princeton: Princeton University Press. ISBN *978-0-691-03705-9*. OCLC *929349643*.

Greiner, W.; Müller, B.; Rafelski, J. (2012). *Quantum Electrodynamics of Strong Fields: With an Introduction into Modern Relativistic Quantum Mechanics*. Springer. *doi*:*10.1007/978-3-642-82272-8*. ISBN *978-0-387-13404-8*. LCCN *84026824*. OCLC *317097176*.

Haisch, B.; Rueda, A.; Puthoff, H. E. (1994). *"Inertia as a Zero-Point-Field Lorentz Force"* (PDF). Physical Review A. 49 (2):

678–694. Bibcode:1994PhRvA..49..678H. doi:10.1103/PhysRevA.49.678. PMID 9910287.

Heisenberg, W.; Euler, H. (1936). "Folgerungen aus der Diracschen Theorie des Positrons". Zeitschrift für Physik. 98 (11–12): 714–732. arXiv:physics/0605038. Bibcode:1936ZPhy...98..714H. doi:10.1007/BF01343663. ISSN 1434-6001.

Heitler, W. (1984). *The Quantum Theory of Radiation* (1954 reprint 3rd ed.). New York: Dover Publications. ISBN 978-0486645582. LCCN 83005201. OCLC 924845769.

Heyl, J. S.; Shaviv, N. J. (2000). "Polarization evolution in strong magnetic fields". Monthly Notices of the Royal Astronomical Society. 311 (3): 555–564. arXiv:astro-ph/9909339. Bibcode:2000MNRAS.311..555H. doi:10.1046/j.1365-8711.2000.03076.x. ISSN 0035-8711.

Itzykson, C.; Zuber, J.-B. (1980). *Quantum Field Theory* (2005 ed.). Mineola, New York: Dover Publications. ISBN 978-0486445687. LCCN 2005053026. OCLC 61200849.

Kostelecký, V. Alan; Mewes, M. (2009). "Electrodynamics with Lorentz-violating operators of arbitrary dimension". Physical Review D. 80 (1): 015020. arXiv:0905.0031. Bibcode:2009PhRvD..80a5020K. doi:10.1103/PhysRevD.80.015020. ISSN 1550-7998.

Kostelecký, V. Alan; Mewes, M. (2013). "Constraints on Relativity Violations from Gamma-Ray Bursts". Physical Review Letters. 110 (20): 201601. arXiv:1301.5367. Bibcode:2013PhRvL.110t1601K. doi:10.1103/PhysRevLett.110.201601. ISSN 0031-9007. PMID 25167393.

Kragh, H. (2012). "Preludes to Dark Energy: Zero-Point Energy and Vacuum Speculations". Archive for History of Exact Sciences. 66 (3): 199–240. *arXiv*:*1111.4623*. *doi*:*10.1007/s00407-011-0092-3*. *ISSN* *0003-9519*.

Kragh, H. S.; Overduin, J. M. (2014). *The Weight of the Vacuum : A Scientific History of Dark Energy*. New York: Springer. *ISBN* *978-3-642-55089-8*. *LCCN* *2014938218*. *OCLC* *884863929*.

Kuhn, T. (1978). *Black-Body Theory and the Quantum Discontinuity, 1894-1912*. New York: Oxford University Press. *ISBN* *978-0-19-502383-1*. *LCCN* *77019022*. *OCLC* *803538583*.

Lahteenmaki, P.; Paraoanu, G. S.; Hassel, J.; Hakonen, P. J. (2013). "Dynamical Casimir Effect in a Josephson Metamaterial". Proceedings of the National Academy of Sciences. 110 (11): 4234–4238. *arXiv*:*1111.5608*. *Bibcode*:*2013PNAS..110.4234L*. *doi*:*10.1073/pnas.1212705110*. *ISSN* *0027-8424*.

Leuchs, G.; Sánchez-Soto, L. L. (2013). "A Sum Rule For Charged Elementary Particles". The European Physical Journal D. 67 (3): 57. *arXiv*:*1301.3923*. *Bibcode*:*2013EPJD...67...57L*. *doi*:*10.1140/epjd/e2013-30577-8*. *ISSN* *1434-6060*.

Loudon, R. (2000). *The Quantum Theory of Light* (3rd ed.). Oxford: Oxford University Press. *ISBN* *978-0198501770*. *LCCN* *2001265846*. *OCLC* *44602993*.

Mignani, R. P.; Testa, V.; González Caniulef, D.; Taverna, R.; Turolla, R.; Zane, S.; Wu, K. (2017). *"Evidence for vacuum birefringence from the first optical-polarimetry measurement of the isolated neutron star RX J1856.5−3754"* (PDF). Monthly Notices of the Royal Astronomical Society. 465 (1): 492–500.

arXiv:*1610.08323*. *Bibcode*:*2017MNRAS.465..492M*. *doi*:*10.1093/mnras/stw2798*. *ISSN 0035-8711*.

Milonni, P. W. (1994). *The Quantum Vacuum: An Introduction to Quantum Electrodynamics*. Boston: Academic Press. *ISBN 978-0124980808*. *LCCN 93029780*. *OCLC 422797902*.
Milonni, P. W. (2009). *"Zero-Point Energy"*. In Greenberger, Daniel; Hentschel, Klaus; Weinert, Friedel (eds.). Compendium of Quantum Physics: Concepts, Experiments, History and Philosophy. In D. Greenberger, K. Hentschel and F. Wienert (Eds.), Compendium of Quantum Physics Concepts, Experiments, History and Philosophy (Pp.). Berlin, Heidelberg: Springer. pp. 864–866. *arXiv*:*0811.2516*. *doi*:*10.1007/978-3-540-70626-7*. *ISBN 9783540706229*. *LCCN 2008942038*. *OCLC 297803628*.

Peebles, P. J. E.; *Ratra, Bharat* (2003). "The Cosmological Constant and Dark Energy". Reviews of Modern Physics. 75 (2): 559–606. *arXiv*:*astro-ph/0207347*. *Bibcode*:*2003RvMP...75..559P*. *doi*:*10.1103/RevModPhys.75.559*. *ISSN 0034-6861*.

Power, E. A. (1964). Introductory Quantum Electrodynamics. London: Longmans. *LCCN 65020006*. *OCLC 490279969*.

Rafelski, J.; Muller, B. (1985). *Structured Vacuum: Thinking About Nothing* (PDF). H. Deutsch: Thun. *ISBN 978-3871448898*. *LCCN 86175968*. *OCLC 946050522*.

Rees, Martin, ed. (2012). *Universe*. New York: DK Pub. *ISBN 978-0-7566-9841-6*. *LCCN 2011277855*. *OCLC 851193468*.

Riek, C.; Seletskiy, D. V.; Moskalenko, A. S.; Schmidt, J. F.; Krauspe, P.; Eckart, S.; Eggert, S.; Burkard, G.; Leitenstorfer, A. (2015). *"Direct Sampling of Electric-Field Vacuum Fluctuations"* (PDF). Science. 350 (6259): 420–423.

Bibcode:*2015Sci...350..420R*. doi:*10.1126/science.aac9788*. ISSN *0036-8075*. PMID *26429882*.

Rugh, S. E.; Zinkernagel, H. (2002). "The Quantum Vacuum and the Cosmological Constant Problem". Studies in History and Philosophy of Science Part B: Studies in History and Philosophy of Modern Physics. 33 (4): 663–705. *arXiv:hep-th/0012253*. Bibcode:*2002SHPMP..33..663R*. doi:*10.1016/S1355-2198(02)00033-3*. ISSN *1355-2198*.

Schwinger, J. (1998a). Particles, Sources, and Fields: Volume I. Reading, Massachusetts: Advanced Book Program, Perseus Books. ISBN *978-0-7382-0053-8*. LCCN *98087896*. OCLC *40544377*.

Schwinger, J. (1998b). Particles, Sources, and Fields: Volume II. Reading, Massachusetts: Advanced Book Program, Perseus Books. ISBN *978-0-7382-0054-5*. LCCN *98087896*. OCLC *40544377*.

Schwinger, J. (1998c). Particles, Sources, and Fields: Volume III. Reading, Massachusetts: Advanced Book Program, Perseus Books. ISBN *978-0-7382-0055-2*. LCCN *98087896*. OCLC *40544377*.

Sciama, D. W. (1991). "The Physical Significance of the Vacuum State of a Quantum Field". In *Saunders, Simon*; *Brown, Harvey R.* (eds.). The Philosophy of Vacuum. Oxford: Oxford University Press. ISBN *978-0198244493*. LCCN *90048906*. OCLC *774073198*.

Saunders, Simon; *Brown, Harvey R.*, eds. (1991). The Philosophy of Vacuum. Oxford: Oxford University Press. ISBN *978-0198244493*. LCCN *90048906*. OCLC *774073198*.

Urban, M.; Couchot, F.; Sarazin, X.; Djannati-Atai, A. (2013). "The Quantum Vacuum as the Origin of the Speed of Light".

The European Physical Journal D. 67 (3): 58. arXiv:1302.6165. Bibcode:2013EPJD...67...58U. doi:10.1140/epjd/e2013-30578-7. ISSN 1434-6060.

Weinberg, S. (1989). "The Cosmological Constant Problem" (PDF). Reviews of Modern Physics. 61 (1): 1–23. Bibcode:1989RvMP...61....1W. doi:10.1103/RevModPhys.61.1. hdl:2152/61094. ISSN 0034-6861.

Weinberg, S. (2015). Lectures on Quantum Mechanics (2nd ed.). Cambridge: Cambridge University Press. ISBN 978-1-107-11166-0. LCCN 2015021123. OCLC 910664598.

Weisskopf, V. (1936). "Über die Elektrodynamik des Vakuums auf Grund des Quanten-Theorie des Elektrons" (PDF). Kongelige Danske Videnskabernes Selskab, Mathematisk-fysiske Meddelelse. 24 (6): 3–39.

Wilson, C. M.; Johansson, G.; Pourkabirian, A.; Simoen, M.; Johansson, J. R.; Duty, T.; Nori, F.; Delsing, P. (2011). "Observation of the Dynamical Casimir Effect in a Superconducting Circuit". Nature. 479 (7373): 376–379. arXiv:1105.4714. Bibcode:2011Natur.479..376W. doi:10.1038/nature10561. ISSN 0028-0836. PMID 22094697.

Balungi Francis, (2010) "A hypothetical investigation into the realm of the microscopic and macroscopic universes beyond the standard model" general physics arXiv:1002.2287v1 [physics.gen-ph]

Hawking, Stephen (1975). "Particle Creation by Black Holes". Commun. Math. Phys. 43 (3): 199–220. Bibcode:1975CMaPh..43..199H.

Hawking, S. W. (1974). "Black hole explosions?". Nature.248(5443):30–31.

Bibcode:1974Natur.248...30H.doi:10.1038/248030a0.

Carlo Rovelli (2003) "Quantum Gravity" Draft of the Book Pdf
Some few texts used are from Wikipedia
https://creativecommons.org/licenses/by-sa/3.0/
D. N. Page, Phys. Rev. D 13, 198 (1976).

C. Gao and Y.Lu, Pulsations of a black hole in LQG (2012) arXiv:1706.08009v3
A.H. Chamseddine and V.Mukhanov, Non singular black hole (2016) arXiv 1612.05861v1

M.Bojowald and G.M.Paily, A no-singularity scenario in LQG (2012) arXiv: 1206.5765v1

P.Singh, class.Quant.Grav,26,125005(2009), arXiv:0901.2750

P.Singh and F.Vidotto, Phys.Rev, D83,064027(2011) arXiv:1012.1307

C.Rovelli and F.Vidotto, Phy. Rev,111(9) 091303(2013) arXiv:1307.3228v2

M.Bojowald, Initial conditions for a universe, Gravity Research Foundation (2003)

A.Ashtekar, Singularity Resolution in Loop Quantum Cosmology (2008) arXiv:0812.4703v1

J.Brunneumann and T.Thiemann, On singularity avoidance in Loop Quantum Gravity (2005) arXiv:0505032v1

L.Modesto, Disappearence of the Black hole singularity in Quantum gravity (2004) arXiv:0407097v2

Mikhailov, A.A. (1959).Mon. Not. Roy. Astron. Soc.,119, 593.

P. Merat etal.(1974). Astron & Astrophys 32, 471-475

Trempler, R.J. (1956).Helv. Phys. Acta, Suppl.,IV, 106.

Trempler, R.J. (1932). " The deflection of light in the sun's gravitational field "Astronomical society of the pacific 167

Einstein, A. (1916).Ann. d. Phys.,49, 769; (1923).The Principle of Relativity, (translators Perret, W. and Jeffery, G.B.), (Dover Publications, Inc., New York), pp. 109–164.

Von Klüber, H. (1960). InVistas in Astronomy, Vol. 3, pp. 47–77.

K. Hentschel (1992). Erwin Finlay Freundlich and testing Einstein theory of relativity, Communicated by J.D. North
Muhleman, D.O., Ekers, R.D. and Fomalont, E.B. (1970).Phys. Rev. Lett.,24, 1377

Mikhailov, A.A. (1956).Astron. Zh.,33, 912.

Dyson, F.W., Eddington, A.S. and Davidson, C. (1920).Phil. Trans. Roy. sog., A220, 291

Chant, C.A. and Young, R.K. (1924).Publ. Dom. Astron. Obs.,2, 275.

Campbell, W.W. and Trumbler, R.J. (1923).Lick Obs. Bull.,11, 41.

Freundlich, E.F., von Klüber, H. and von Brunn, A. (1931).Abhandl. Preuss. Akad. Wiss. Berlin, Phys. Math. Kl., No.l;Z. Astrophys.,3, 171

Mikhailov, A.A. (1949).Expeditions to Observe the Total Solar Eclipse of 21 September, 1941, (report), (ed. Fesenkov, V.G.), (Publications of the Academy of Sciences, U.S.S.R.), pp. 337–351.

S.P. Martin, in Perspectives on Supersymmetry , edited by G.L. Kane (World Scientific, Singapore, 1998) pp. 1–98; and a longer archive version in hep-ph/9709356; I.J.R. Aitchison, hep-ph/0505105.

Mara Beller, Quantum Dialogue: The Making of a Revolution. University of Chicago Press, Chicago, 2001.

Morrison, Philp: "The Neutrino, scientific American, Vol 194,no.1 (1956),pp.58-68.
R. Haag, J. T. Lopuszanski and M. Sohnius, Nucl. Phys. B88, 257 (1975) S.R. Coleman and J. Mandula, Phys.Rev. 159 (1967) 1251.

H.P. Nilles, Phys. Reports 110, 1 (1984).

P. Nath, R. Arnowitt, and A.H. Chamseddine, Applied $N = 1$ Supergravity (World Scientific, Singapore, 1984).

S.P. Martin, in Perspectives on Supersymmetry , edited by G.L. Kane (World Scientific, Singapore, 1998) pp. 1–98; and a longer archive version in hep-ph/9709356; I.J.R. Aitchison, hep-ph/0505105.

S. Weinberg, The Quantum Theory of Fields, VolumeIII: Supersymmetry (Cambridge University Press, Cambridge,UK, 2000).

E. Witten, Nucl. Phys. B188, 513 (1981).

S. Dimopoulos and H. Georgi, Nucl. Phys. B193, 150(1981).

N. Sakai, Z. Phys. C11, 153 (1981);R.K. Kaul, Phys. Lett. 109B, 19 (1982).

L. Susskind, Phys. Reports 104, 181 (1984).
L. Girardello and M. Grisaru, Nucl. Phys. B194, 65(1982); L.J. Hall and L. Randall,

Phys. Rev. Lett. 65, 2939(1990);I. Jack and D.R.T. Jones, Phys. Lett. B457, 101 (1999).

For a review, see N. Polonsky, Supersymmetry: Structureand phenomena. Extensions of the standard model, Lect.Notes Phys. M68, 1 (2001).

G. Bertone, D. Hooper and J. Silk, Phys. Reports 405, 279 (2005).

G. Jungman, M. Kamionkowski, and K. Griest, Phys. Reports 267, 195 (1996).

V. Agrawal, S.M. Barr, J.F. Donoghue and D. Seckel,Phys. Rev. D57, 5480 (1998).

N. Arkani-Hamed and S. Dimopoulos, JHEP 0506, 073(2005); G.F. Giudice and A. Romanino, Nucl. Phys. B699, 65(2004) [erratum: B706, 65 (2005)]. July 27, 2006 11:28

en.wikipedia.org/wiki/Supersymmetry - 52k - Cached - Similar pages
en.wikipedia.org/wiki/Grand_unification_theory - 39k - Cached - Similar pages

In cosmology, the Planck epoch (or Planck era), named after Max Planck, is the earliest period of time in the history of the universe, en.wikipedia.org/wiki/**Planck_epoch** - 23k - Cached - Similar pages

L. Shapiro and J. Sol`a, Phys. Lett. B 530, 10 (2002);

E. V.Gorbar and I. L. Shapiro, JHEP 02, 021 (2003); A. M. Pelinson,

L. Shapiro, and F. I. Takakura, Nucl. Phys. B 648, 417 (2003).

A. Starobinsky, Phys. Lett. B 91, 99 (1980).

G. F. R. Ellis, J. Murugan, and C. G. Tsagas, Class. Quant. Grav.21, 233 (2004).

H. V. Peiris et al., Astrophys. J. Suppl. 148, 213 (2003).

D. N. Spergel et al., astro-ph/0603449.

Vilenkin, Phys. Rev. D 32, 2511 (1985).

A. Starobinsky, Pis'ma Astron. Zh 9, 579 (1983).

A.H. Guth, Phys. Rev. D23, 347 (1981).

A.D. Linde, Phys. Lett. B108, 389 (1982); A. Albrecht, P.J.

Steinhardt, Phys.Rev. Lett. 48, 1220 (1982).

A.D. Linde, Phys Lett. B129, 177 (1983).

N. Makino, M. Sasaki, Prog. Theor. Phys. 86, 103 (1991);

D. Kaiser, Phys. Rev.D52, 4295 (1995).

H. Goldberg, Phys. Rev. Lett. 50, 1419 (1983).

E. Kolb and M. Turner, The Early Universe (Addison-Wesley, Reading, MA,1990).

W. Garretson and E. Carlson, Phys. Lett. B 315, 232(1993); H. Goldberg, hep-ph/0003197.

Eddington, A. S., The Internal Constitution of the Stars (Cambridge University Press, England,1926), p. 16

Duncan R .C. & Thompson C., Ap.J.392, L 9 (1992). Thompson , C, Duncan , R .C ., Woods , P., Kouveliotou , C ., Finger , M.H. & van Parad ij s , J .,ApJ, submitted , astro-ph /9908086, (2000).

Schwinger, J.,Phys. Rev.73, 416L (1948)

Carlip, S.: Quantum gravity: a progress report. Rept. Prog. Phys. 64, 885 (2001).arXiv:gr-qc/0108040

Kerr,R.P.: Gravitational field of a spinning mass as an example of algebraically special metrics.

Phys. Rev. Lett. 11, 237–238 (1963)

Bekenstein, J.: Black holes and the second law. Lett. Nuovo Cim. 4, 737–740 (1972)

Bardeen, J.M., Carter, B., Hawking, S.: The four laws of black hole mechanics. Commun.

Math. Phys. 31, 161–170 (1973)

Tolman, R.: Relativity, Thermodynamics, and Cosmology. Dover Books on Physics Series.

Dover Publications, New York (1987)
Oppenheimer, J., Volkoff, G.: On massive neutron cores. Phys. Rev. 55, 374–381 (1939)

Tolman, R.C.: Static solutions of einstein's field equations for spheres of fluid, Phys. Rev. 55,364–373 (1939)

Zel'dovich Y.B.: Zh. Eksp. Teoret. Fiz.41, 1609 (1961)

Bondi, H.: Massive spheres in general relativity. Proc. Roy. Soc. Lond. A281, 303–317 (1964)

Sorkin, R.D., Wald, R.M., Zhang, Z.J.: Entropy of selfgravitating radiation. Gen. Rel. Grav. 1127–1146 (1981)

Newman, E.T., Couch, R., Chinnapared, K., Exton, A., Prakash, A., et al.: Metric of a rotating,charged mass. J. Math. Phys. 6, 918–919 (1965)

Ginzburg, V., Ozernoi, L.: Sov. Phys. JETP 20, 689 (1965)

Doroshkevich, A., Zel'dovich, Y., Novikov I.: Gravitational collapse of nonsymmetric and rotating masses, JETP 49 (1965)

Israel, W.: Event horizons in static vacuum space-times. Phys. Rev. 164, 1776–1779 (1967)
Israel,W.: Event horizons in static electrovac space-times. Commun. Math. Phys. 8, 245–260 (1968)

Loop quantum gravity does provide such a prediction [363, 364], and it disagrees with the semiclassical

Carter, B.: Axisymmetric black hole has only two degrees of freedom. Phys. Rev. Lett. 26, 331–333(1971)

Penrose, R.: Gravitational collapse: the role of general relativity. Riv. Nuovo Cim. 1, 252–276 (1969)

Christodoulou, D.: Reversible and irreversible transformations in black hole physics. Phys. Rev. Lett. 25, 1596–1597 (1970)

Christodoulou, D., Ruffini, R.: Reversible transformations of a charged black hole. Phys. Rev. D4, 3552–3555 (1971)

Hawking, S.: Particle creation by black holes. Commun. Math. Phys. 43, 199–220 (1975)

Klein, O.: Die reflexion von elektronen an einem potential sprung nach der relativistischen dynamik von dirac. Z. Phys. 53, 157 (1929)

Wald, R.M.: General Relativity. The University of Chicago Press, Chicago (1984)

Hawking, S.W.: Black hole explosions. Nature 248, 30–31 (1974)

Hawking, S., Ellis, G.: The large scale structure of space-time. Cambridge University Press, Cambridge (1973)

Carter, B.: Black hole equilibrium states, In Black Holes—Les astres occlus. Gordon and Breach Science Publishers, (1973)

Hawking, S.W.: Gravitational radiation from colliding black holes. Phys. Rev. Lett. 26, 1344–1346 (1971)

Hawking, S.: Black holes in general relativity. Commun. Math. Phys. 25, 152–166 (1972)

Bekenstein, J.: Extraction of energy and charge from a black hole. Phys. Rev. D7, 949–953 (1973)

Bekenstein, J.D.: Black holes and entropy. Phys. Rev. D7, 2333–2346 (1973)

Hawking, S.: Quantum gravity and path integrals. Phys. Rev. D18, 1747–1753 (1978)
Gross, D.J., Perry, M.J., Yaffe, L.G.: Instability of flat space at finite temperature. Phys. Rev. D25, 330–355 (1982)

Unruh, W.G., Wald, R.M.: What happens when an accelerating observer detects a rindler particle. Phys. Rev. D29, 1047–1056 (1984)

Bekenstein, J.D.: Auniversal upper bound on the entropy to energy ratio for bounded systems. Phys. Rev. D23, 287 (1981)

Unruh,W.,Wald, R.M.: Acceleration radiation and generalized second law of thermodynamics. Phys. Rev. D25, 942–958 (1982)

Unruh, W., Wald, R.M.: Entropy bounds, acceleration radiation, and the generalized second law. Phys. Rev. D27, 2271–2276 (1983)

Image : MPI for gravitational physics / W.Benger-zib

Tomilin,K.A., (1999). "Natural Systems Of Units: To The Centenary Aniniversary Of   The Planck Systems", 287-296

Sivaram, C. (2007). "What Is Special About the Planck Mass"? arXiv:0707.0058v1

H. Georgi and S.L. Glahow. (1974) "Unity Of All Elementary-Particle Forces". Phys. Rev. Letters 32, 438 Luigi Maxmilian Caligiuri, Amrit Sorli. Gravity Originates from Variable Energy Density of Quantum Vacuum. American Journal of Modern Physics. Vol. 3, No. 3, 2014, pp. 118-128. doi: 10.11648/j.ajmp.20140303.11

Philip J. Tattersall,(2018) Quantum Vacuum Energy and the Emergence of Gravity. doi:10.5539/apr.v10n2p1

H. E. Puthoff (1989) Gravity as a zero-point-fluctuation force PHYSICAL REVIEW A VOLUME 39, NUMBER 5

Balungi Francis (2018) "Quantum Gravity in a Nutshell1" Book.

E.Verlinde (2016) Emergent Gravity and the Dark Universe, arXiv:1611.02269v2[hep-th]

S.Hossenfelder (2018) The Redshift-Dependence of Radial Acceleration: Modified gravity versus particle dark matter, arXiv:1803.08683v1[gr-qc]

Robert J. Scherrer (2004) Purely kinetic k-essence as unified dark matter, arXiv:astro-ph/0402316v3

J.S.Farnes (2018), Aunifying theory of dark energy and dark matter: Negative masses and matter creation within a modified ΛCDM framework, arXiv:1712.07962v2[physics.gen-ph]

Gustav M Obermair (2013), Primordial Planck mass black holes (PPMBHs) as candidates for dark matter? J. Phys:conf.Ser.442012066

V.Cooray etal...(2017), An alternative approach to estimate the vacuum energy density of free space, doi:10.20944/preprints201707.0048.v1

M.Milgrom, (1983) A modification of the Newtonian dynamics: Implications for galaxies, Astrophys.J.270, 371.

Acknowledgments

This book would never have been completed without the patience and dedication of my wife, Wanyana Ritah. She performed the wonderful and difficult task of editing major parts of the book and helped in researching many details necessary to complete it.

I wish to thank several colleagues for their help and extensive comments on the manuscript. I particularly thank a total of 200 online physics friends and SUSP science foundation members, for a careful reading of the manuscript. Many graduate students have contributed over the years to developing my Quantum theory of gravity.

I also wish to thank my editors, at SUSP science Foundation and Bill stone Services for their enthusiasm and support. Finally, I thank our family for their patience, love, and support during the four years of working on this book.

About the Author

Balungi Francis was born in Kampala, Uganda, to a single poor mother, grew up in Kawempe, and later joined Makerere Universty in 2006, graduating with a Bachelor Science degree in Land Surveying in 2010. For four years he taught in Kampala City high schools, majoring in the fields of Gravitation and Quantum Physics. His first book, "Mathematical Foundation of the Quantum theory of Gravity," won the Young Kampala Innovative Prize and was mentioned in the African Next Einstein Book Prize (ANE).

He has spent over 15years researching and discovering connections in physics, mathematics, geometry, cosmology, quantum mechanics, gravity, in addition to astrophysics, unified physics and geographical information systems . These studies led to his groundbreaking theories, published papers, books and patented inventions in the science of Quantum Gravity, which have received worldwide recognition.

From these discoveries, Balungi founded the SUSP (Solutions to the Unsolved Scientific Problems) Project Foundation in 2004 - now known as the SUSP Science Foundation. As its current Director of Research, Balungi leads physicists, mathematicians and engineers in exploring Quantum Gravity principles and their implications in our world today and for future generations.

Balungi launched the Visionary School of Quantum Gravity in 2016 in order to bring the learning and community further together. It's the first and only Quantum Gravity physics program of its kind, educating thousands of students from over 80 countries.

The book "Quantum Gravity in a Nutshell1", a most recommend book in quantum gravity research, was produced based on Balungi's discoveries and their potential for generations to come. Balungi is currently guiding the Foundation, speaking to audiences worldwide, and continuing his groundbreaking research.

Contact Balungi Francis at the following address:

Email: balungif@gmail.com
bfrancis@cedat.mak.ac.ug

Mobile Telephone: +256703683756

Made in the USA
Columbia, SC
01 August 2021